Add Me

Your Guide Book To Making New Friends via Facebook and its Advantages

Author
Rehan Allahwala

My sincere thanks to everyone who supported and contributed in making the informative booklet.

www.**RehanAllahwala**.com

Title:

Add Me: Your Guide Book To Making New Friends via Facebook and its Advantages

ISBN-13: 978-1523725144

ISBN-10: 1523725141

Author:

Rehan Allahwala

Copyright Holder:

Publisher:

Amazon.com

Synopsis:

This book contains some of my methods on how to turn strangers from all over the world into friends, quickly. This method eliminates fears, preconceptions and some old precautions we might have been taught...

Country Restrictions:

No restrictions

Foreword

Meeting new people can be a very rewarding experience. In the past, we had limited tools to make friends and on some occasions, we were taught not to talk to strangers either. Technology and tools have advanced to such a degree, that the whole paradigm of communication is altered forever. We can now immediately connect to people from all over the world. One very effective communication tool that I have been using is Facebook. This book contains some of my methods on how to turn strangers from all over the world into friends, quickly. This method eliminates fears, preconceptions and some old precautions we might have been taught. We can now be friends with people from all over the world in a matter of months, weeks and even minutes. In this book I share my experience in this arena so far so you can do the same.

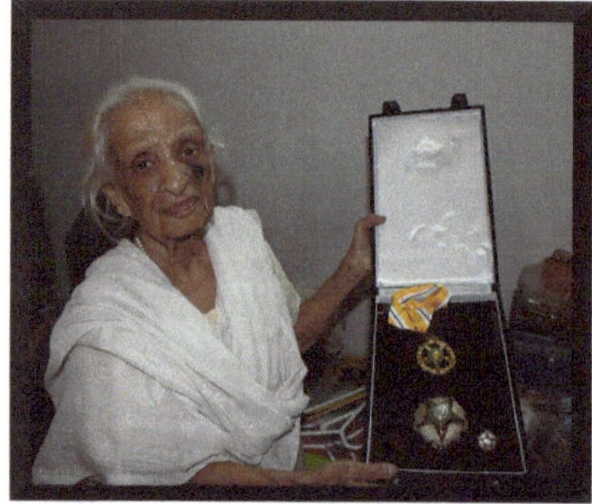

Late Fatima Surraiya Bajia 1930 - 2016

Pride of Performance, Hilal-i-Imtiaz (Pakistan), Order of the Rising Sun (Japan)

Renowned Urdu novelist, playwright and drama writer

Dedication

I dedicate this book to

Late Fatima Surayya Bajia,

who is one of whom I learnt the

best way, how to talk to strangers:

She said - **"They are not strangers rather than one of your own, who you have just not met yet".**

You can learn more about her on

www.bajia.org

~Rehan Allahwala

Table of Contents

1. Introduction

"When we come to this world, we come alone, and we leave alone" said Bajia. However on this amazing journey, we meet a lot of people, some are family, and others are our adopted family, who we call our friends.

Making friends has been a challenge for me most of my life. I was fascinated by people, who were able to go and talk to anyone. Then I met some people who managed to teach me how this can be done. Here is what I know, and I am sharing it with you.

We have been told that talking to strangers is a bad thing and not to interact with many people in our lives, especially if they are different than us. So we develop fear from others. We can get rid of this fear slowly, by doing some things that are of a different pattern than that which we are used to be doing, and slowly this fear goes away.

I think we need to stop telling our children that talking to strangers is a bad thing, rather educate them how to talk and communicate to strangers. This should be a subject introduced and taught in all our schools around the world.

My definition of a stranger is, a person who we have not talked with for 5 minutes yet. Once we do this they become acquaintances and after an hour of conversation or interaction of some kinds, friends.

Meetings strangers with Facebook becomes very interesting, you don't see and meet the person but all of a sudden have a possibility to know so much more about them, what they can't tell you on the first meeting. Facebook made this possible.

Rehan Allahwala in his first office, Pakistan Computers

My journey of interacting with others began when I was 16 and was running Pakistan Computers. I would invite 10 to 40 friends from the computer industry for a lunch party at my home. I introduced them to each other, so this way we developed healthier business relationships. I also did this because I was looking for learning new things and making new friends.

The circle of my friends expended when I got my first email address in 1994. With the introduction of IRC and ICQ (messenger programs) my friendship circle grew even bigger. Now I had real people around the world as friends who I could talk to in real time.

In 2007, when Facebook came out, everything was brought together in the same place: text, photos and videos were all in one place. You could learn about people very, very quickly and you had the most

Informative data you need to make friends right there in front of You - If you looked at WHAT THEY LIKED.

Knowing and learning from different people is an amazing journey. It is like watching many life documentaries, or reading many autobiographies, only that it is all happening in real life.

While visiting a friend of mine in St. Petersburg, Florida, United States, I was explaining to someone, how I have been collecting 10 friends from each country of the world for the past many years.

WHY? I wanted to do this, as I wanted to know the ground reality. I wanted to know what is really is happening around the world, unlike what I heard from the news media. This became my excuse to explore new cultures, different ways of living and try different foods. It just enriched my life in ways that I cannot even count any more. I have been successful achieving this in many countries and yet many more to come.

With this book I like to expose my friends who don't use Facebook yet, to the world what Facebook can bring them. I assigned my newly joined Facebook friends to a group where they can get help, see photos of other people and let them decide if they want to connect to these wonderful people around the world. That is what this booklet is about.

I started this book in 2014 and I hope that we'll publish more versions of it in the future, so that more and more can be added in this booklet and eventually we will not really need this booklet, as the ideas here are working in real life and even transformed further.

I hope you will enjoy this booklet and take full advantage of it. I hope this will allow you to be more open, more accepting and eventually to realize that what we think the truth is may not be complete yet. Only when seeing the complete picture can you decide what is true or false. Facebook allows you to see the complete picture.

2. How It All Started

How I Found My Friends on Facebook

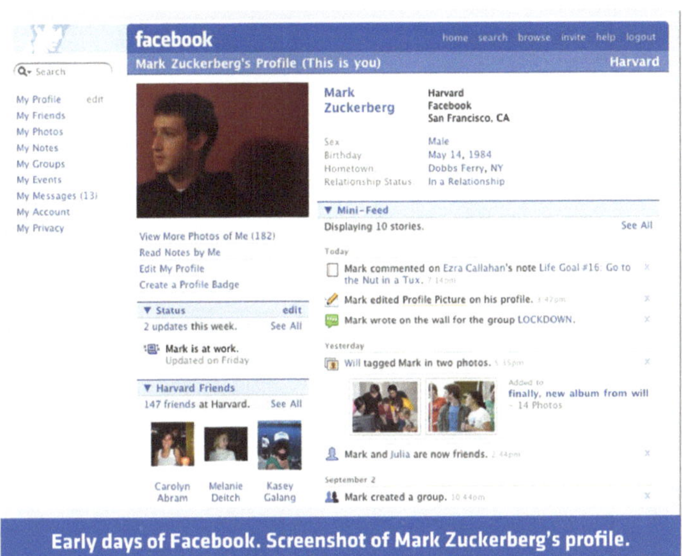

Early days of Facebook. Screenshot of Mark Zuckerberg's profile.

I signed up on Facebook as soon as it was available. At that time Facebook didn't look like what it looks like today. Facebook initially was more about who could post the most attention getting pictures. Since pictures were not my thing so I just left Facebook there. With the passage of time, Facebook changed to where you could post a status and photos, as well as make friends. Now I started using it more and more. Initially out of my friends who I knew it was only Suzanne Bowen and a few others who were active on Facebook.

Slowly and gradually, I started spending more time on Facebook. At the time of ICQ, I would make friends in the places I wanted to know about or would travel to, so I could get first hand information about these places and their cultures. I also wanted to have a friend to greet me at the airport when I would arrived.

Initially it was not easy to make friends. So I found that making friends in many countries were more difficult, but people would always want to help you out. So my line to get a reply from them, or as they say "breaking the ice" would be:

"Hello, I am sorry to interrupt, but can you tell me what is the local time now in your city?"

This sentence would get most replies for me and people would reply to me telling me about it, and it was a great ice breaker. The conversation would go on from there.

My first FB account was closed when I started adding many people. Facebook just closed the account, without giving me a reason. It turned out that the reason was that I was "friending" a lot.

I opened a new account and collected 4000 something friends on it and one fine morning, that also got closed. I was very upset and sad that I lost all these friends who I found over the years, again.

Now Facebook doesn't close your account only suspends you from adding people. Like my mother used to say, **"They PUT YOU IN JAIL".** During this "jail-time", you cannot add people for 30 days.

3. Making Friends On Facebook Successfully

There are few things that you will need to sure that are correct on your Facebook profile before you can pursue this journey. I will explain all these on the upcoming pages so you know what and how to do. These suggestions are very similar to making friends in real life.

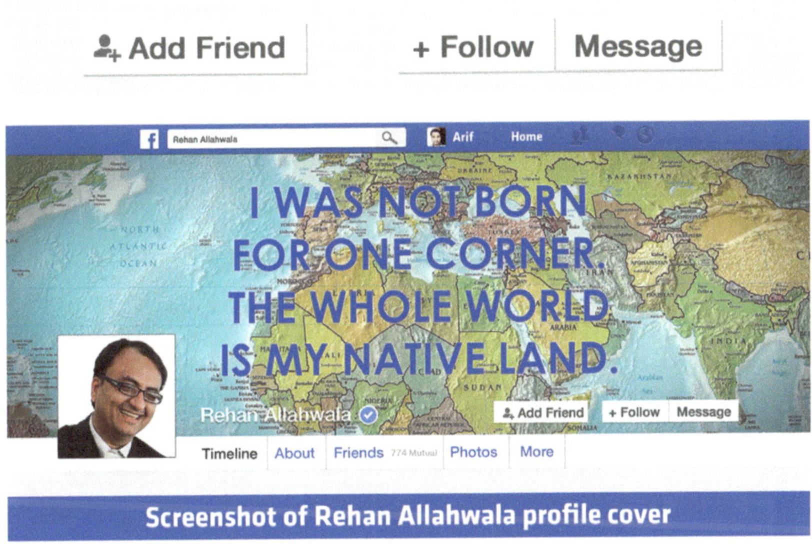

Screenshot of Rehan Allahwala profile cover

After making mistakes, I found the best way to add friends: that is to be honest, clear and direct in asking what you want from your Facebook friend. At this point I would highly recommend reading the book by Dale Carnegie **"How to Win Friends and Influence People"**. This book will help you to create the mind-set for making friends in general, in real life and on FB. You can find it on **www.rehanu.com/friends**

4. How To Prepare Your Facebook Profile For Making Friends

Before you send a friend request you need to take a full step. Just like you would not say: **"Hello, will you be my friend?"** to someone who you just met in person while wearing your night pajamas. Make sure that there is enough collateral, information about you on your Facebook profile for the other person to see who you are and if you are someone they will WANT to be friends with by accepting your friends' request.

These are the things we will fix on your Facebook profile:

a) Profile Photo

b) Cover Photo

c) Your Photos

d) Your information

e) The Follow Function

a) Your Profile Photo:

Your profile photo must be a clear face photo. Choose a photo with smile and showing nothing more but your face, so that the other person can read your face and make a judgment call on it.

sample of a profile photo

Some examples of clear Facebook face photos

I recommend to have at least 5 other face shot's on your profile, which, they are set to "public", so everyone can see it. If someone clicks on your profile photo and they see different faces, they think you are maybe a fake ID. Having five photos of yourself on your profile and ensures others that you're real and that take their fear away.

These are my profile pictures and this is what I recommend for you to have.

Profile Picture and its settings

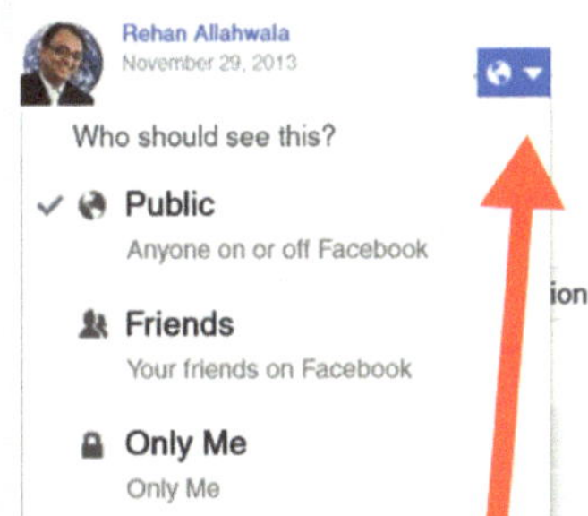

Click on the arrow and choose "Public" from the drop down list.

These are below the examples of my five other profile photos which only show my face.

Five other profile pictures

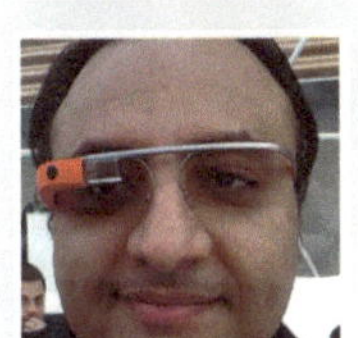

The profile photo has to be your face only, so when you write a message to someone, this small profile picture will be next to your message and you are easily recognizable.

b) Your Cover Photo:

Cover photo is another important part of your profile. These photos will further talk about you: who you are. What your mindset is? And what is your interest in life.

I also recommend having at least five of your cover photos with your friends, who are tagged on this photo.

These are the steps how to tag your friends on the photos:

Step 1:

Move the cursor to the bottom of the photo and click on the "Tag Photo" button.

Step 2: Click on the person in a photo and type his/ her name, then choose from the drop down list.

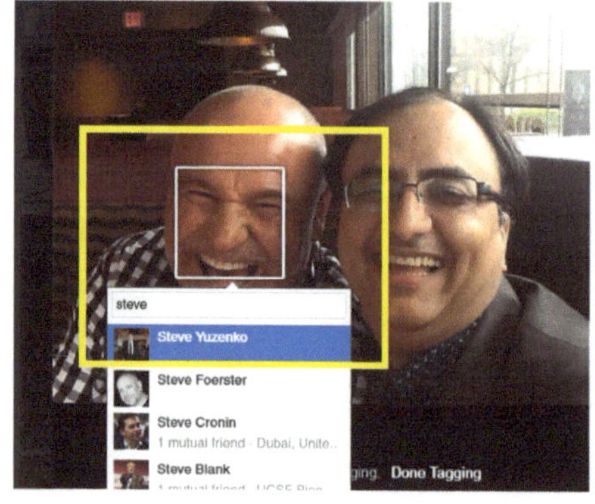

Step 3:

Once you get his/her name from the list, click on "Done Tagging" button.

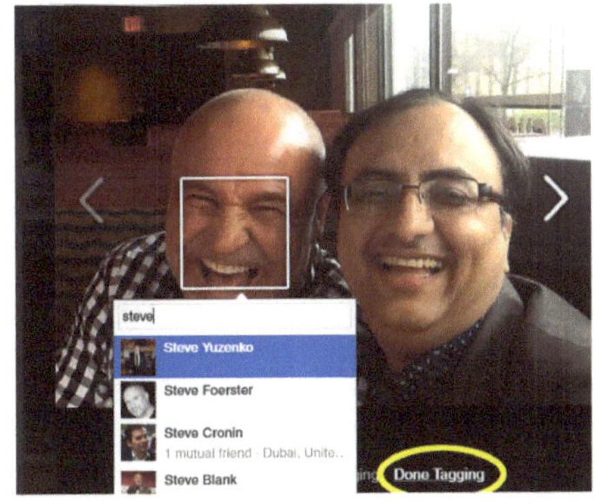

Your cover photo can also be about your interests, nice quotes which you believe in etc., so that the person can see what your thoughts are before considering accepting you as a friend. Keep in mind: friendship here is a lot like in real life. The interaction to make friends will also take time, focus and nurturing.

Sample Cover Links:

www.RehanAllahwala.com/quotes

www.RehanAllahwala.com/wisdom

Cover Photo Samples

Here are a few samples of my cover photos.

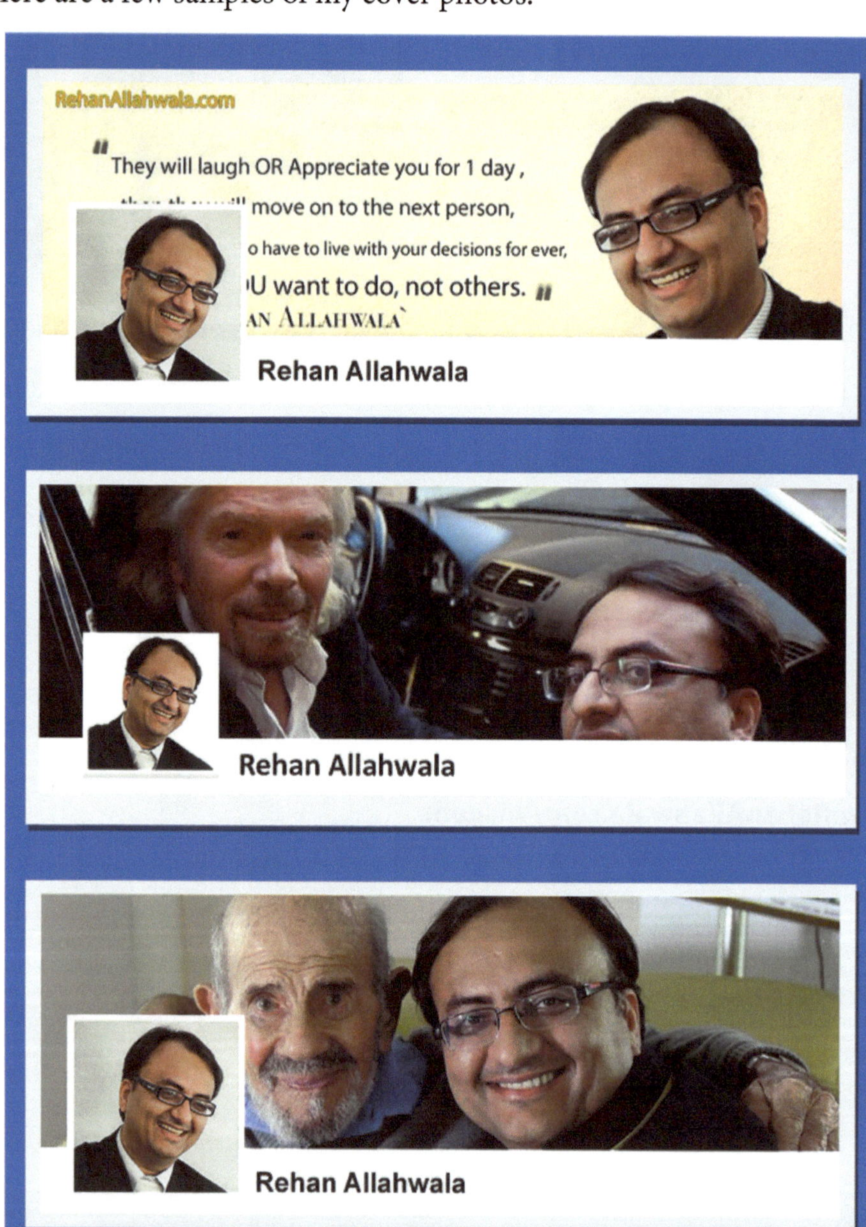

c) Your Photos:

It is also important to have a few photos with people i.e. your co-workers, classmates or friends who you have met and tagged them on different occasions. This also gives more credibility to you and shows who you are, as well as what kind of friends you have. Tagging your friends allow others to see who your friends are as we have all heard the saying that birds are known of their feather flock together; meaning that the company you keep reflects your personality.

People who are in the pictures are tagged with Rehan Allahwala

d) Your Information:

The information page allows you to put different kinds of information about you. This information is about your schooling, your work, your hobbies, interests, music, and books. Add as much information about yourself what you already posted on LinkedIn. If you currently use LinkedIn and have an updated, complete profile, there should be no reason why not to also post it on Facebook too. This information on LinkedIn and Facebook about you saves time for anyone to learn about you without Googling you.

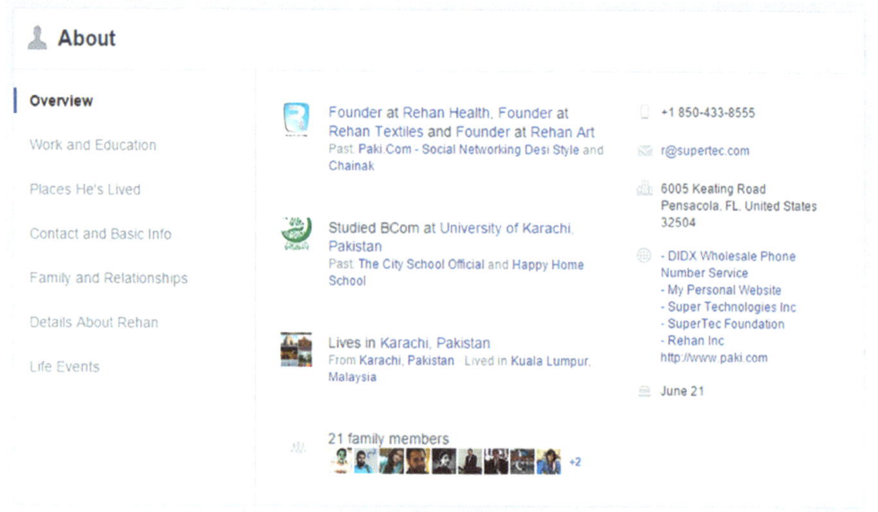

Your information shall include where you studied, what is your specific work in the company. This is also excellent for your professional life, as 75% of the people in the world get a job via someone they know. If your own friends and family do not know what you do, how will they help introducing you to people who can help you grow professionally or get you a better job or work?

e) The Follow Function:

The Facebook follow function is a special feature function of Facebook which allows people to be connected with you without being added as a friend. Being on Facebook is like becoming an ambassador of yourself, your city, your religion, your country and your company. Once anyone on FB, they share a lot of public information, a YouTube video, write about some website, or share a newspaper article, etc. Once you share something on FB, this information CAN be seen by ANYONE, anywhere, anytime and is archived.

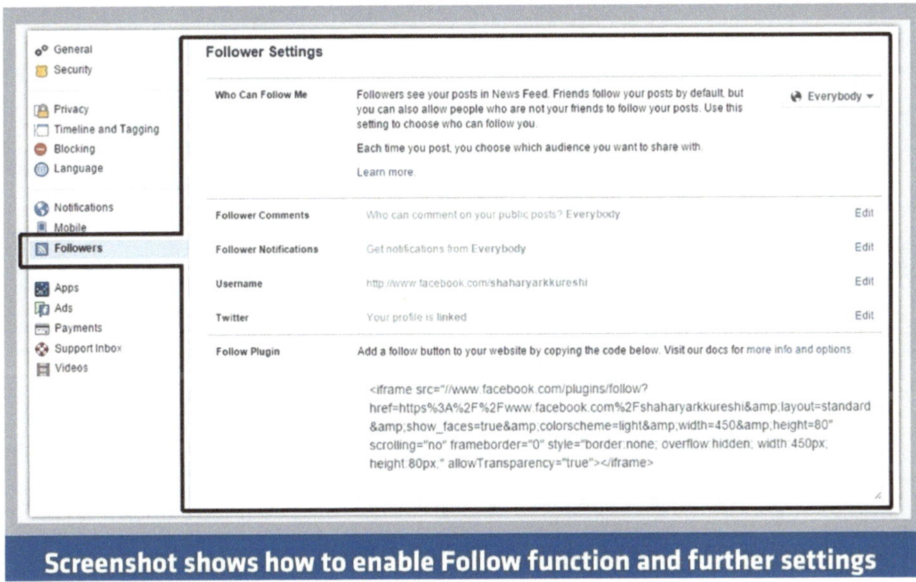

Screenshot shows how to enable Follow function and further settings

Facebook follow function allows you to now have followers like you have on Twitter. Anything you post as PUBLIC, will be available to your your followers and will be sent to them like an RSS Feed (Rich Site Summary) in their news feed. RSS being a follower on someone's FB page is like subscribing to a website and receiving newsletters back on your email. FB follow function will deliver your posts to your followers Newsfeeds.

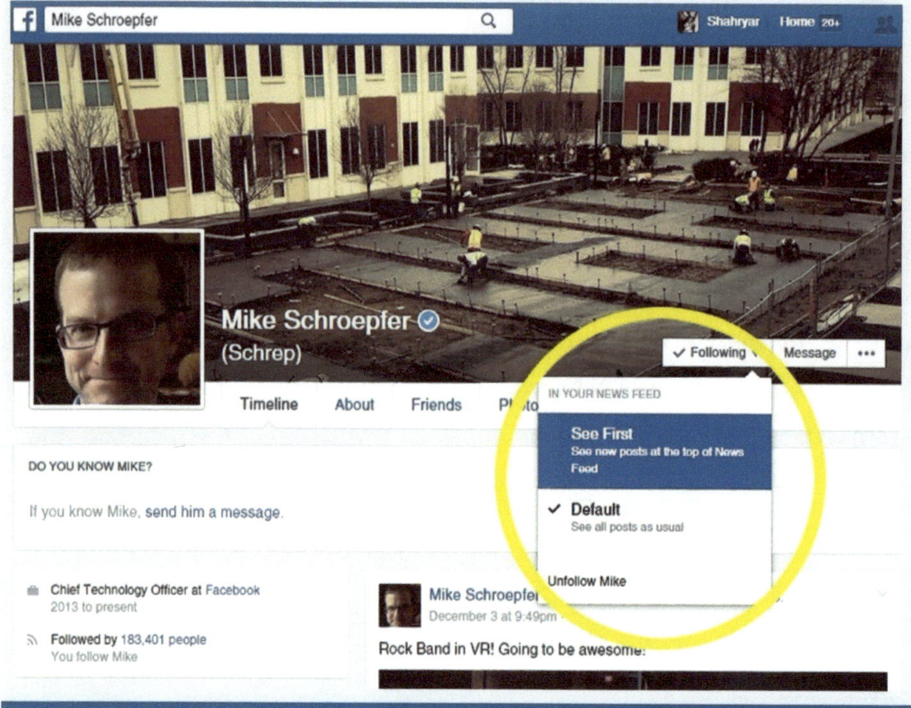

Once enabled, Follow button can be seen on lower right corner of the cover

To open your follow function on laptop, go to **www.facebook.com/about/follow** and set it to PUBLIC.

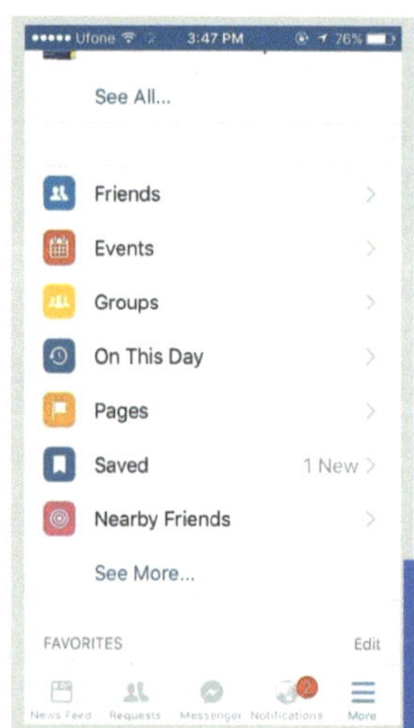

To open follow function on the mobile phone, go to the FB app, and tap on “More” button on bottom right corner of screen. Then go to account settings and follow. You’ll find two options: Friends or Public. You need to tap on the Public option.

Step 1:
Tap on “More” button on bottom right of the screen.

A step-by-step guide about how to enable Follow function in Mobile App

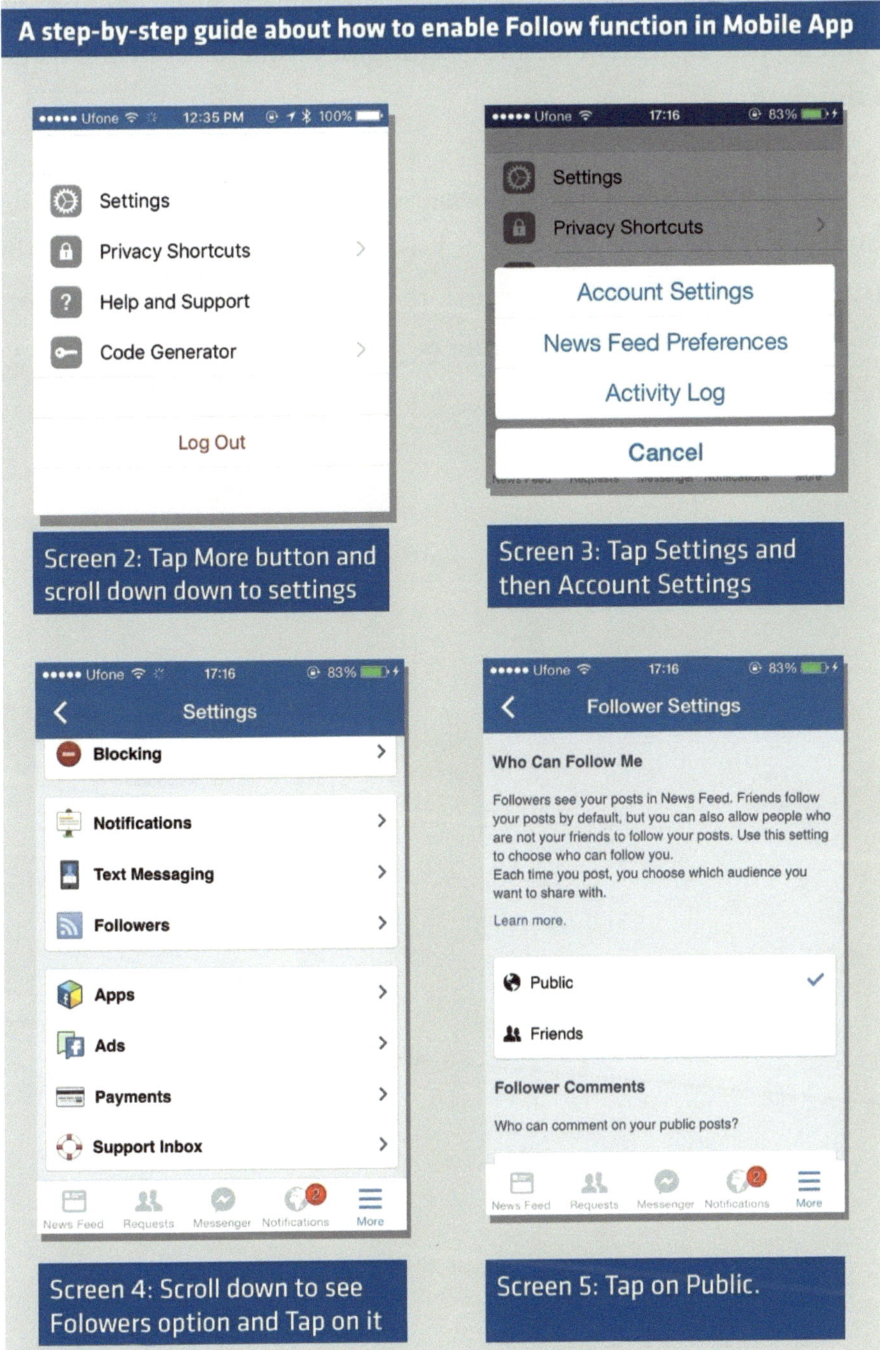

I highly recommend that you open your follower function, even if you do not use Facebook a lot, or do not post a lot of public content at this time.

I recommend this, because people who are friending you will become your followers at least. And in the future when things change for you or career changes, these followers may be of a lot of benefit to you. Facebook has a limit of adding 5000 friends only, but you can have unlimited followers.

5. Benefits Of Adding Friends

Who are the Mutual Friends?

A Mutual Friend is a friend you and another person have in common. Suzanne and I have 500 common friends and these 500 friends are called our "Mutual Friends". These are the people Suzanne and I both know. The rest of the friends on her friend list and my list are our separate friends, who we don't know or share. I am showing you here our Mutual Friends with Shahryar Qureshi, my dear friend.

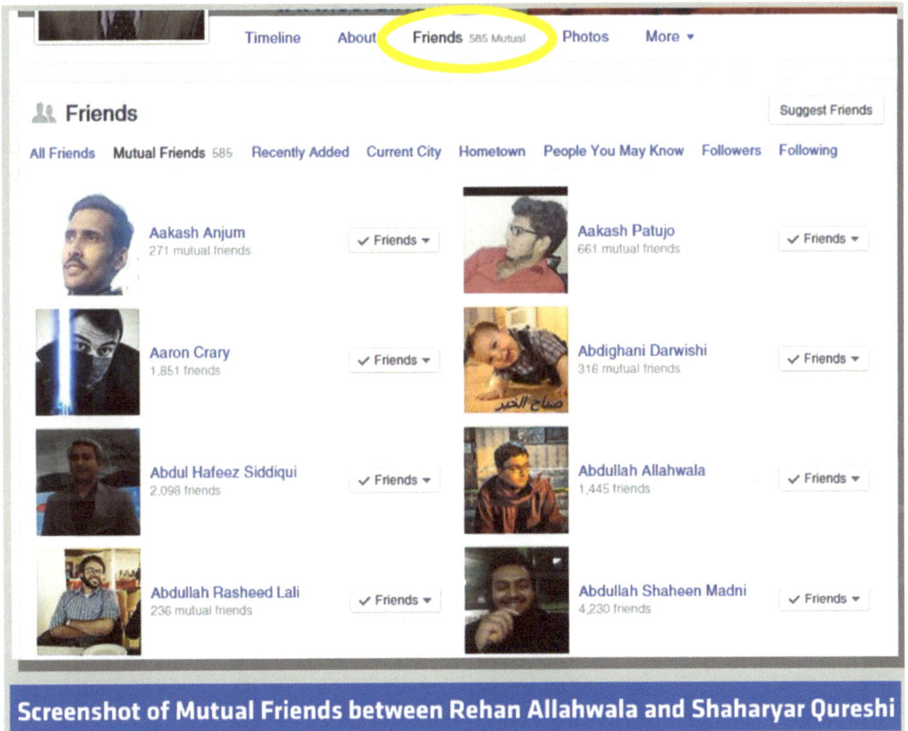

Screenshot of Mutual Friends between Rehan Allahwala and Shaharyar Qureshi

The benefits of Mutual Friends: Building powerful networks

We learn from our Network - people who are around us are our teachers, our friends, our peers, our acquaintances, people we hang around with. They are our teachers and we slowly become like them in admiring their best qualities.

I have spent many years in gathering wonderful friends who are positive, who are learned, who is anyone and everyone.

When you interact with your friends regularly, their goals become your goals and your goals become their goals. Then we will help each other to make these goals, dreams, visions, aspirations to come true. This is the power of network.

The following is my hypothesis on how Mutual Friends with me can help you. So, if you take my friends from my friend list, make them your friends so they are our mutual friends, you will end up learning from this process:

1. How to make new friends
2. How to communicate with strangers and convert them to friends
3. A few words in different languages
4. How people think in other parts of the world
5. How to think "out of the box"
6. The things, which are not available in your country yet, thus bringing you opportunity to bring them to your country
7. Find great teachers
8. Find great students

If you have some of my friends as yours, and have talked to them for 1 hr each only, I guarantee your income, your character and/or your life will improve 500% for sure, potentially all of them.

Can you imagine your life improving 500%? You would like this, wouldn't you?

6. Interesting Friends From Around The World

Here is a list of people who are open to adding random people and engage into conversation with them. You can add them, talk to them and see if this book helped you in convincing these people to add you as a friend.

Bert-Ola Bergstrand

www.facebook.com/bertola.bergstrand

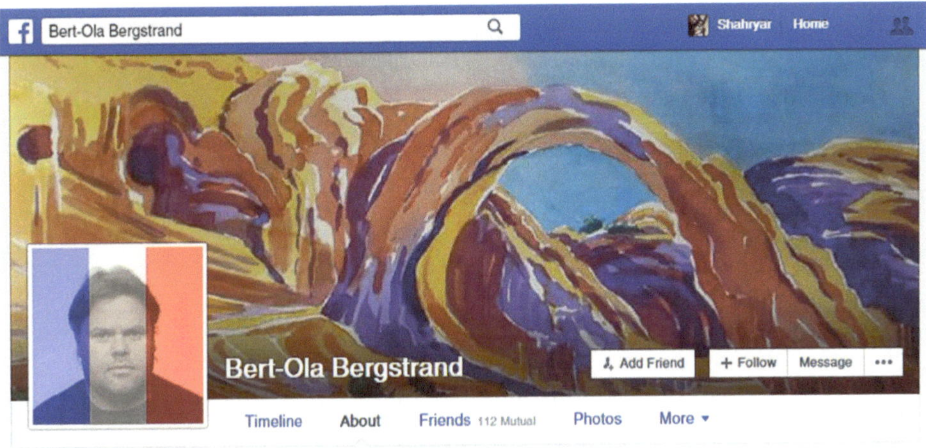

Lisa Canning

www.facebook.com/LisaACanning

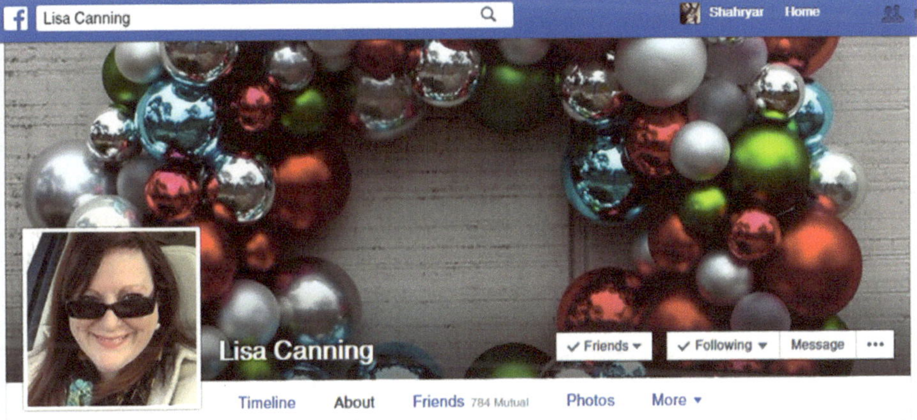

Muhammad Layeeque
www.facebook.com/layeeque86

Pamela Hills
www.facebook.com/pamela.hills.750

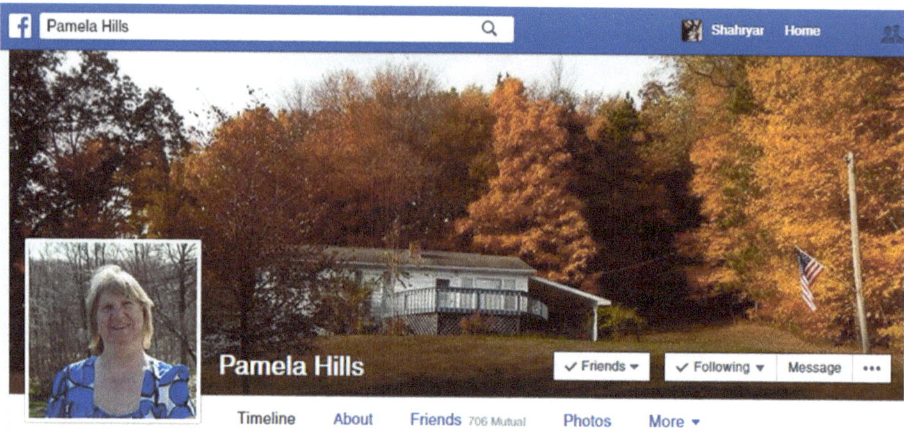

Kaab Ul Ahbar

www.facebook.com/Kaab.Kashi

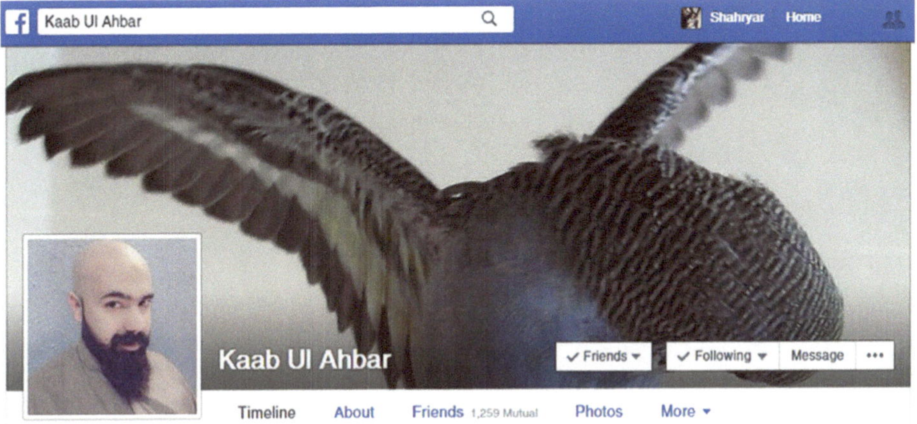

Hammad Ul Hassan

www.facebook.com/hammaduh

7. People You Must Follow

Barack Obama
www.facebook.com/BarackObama

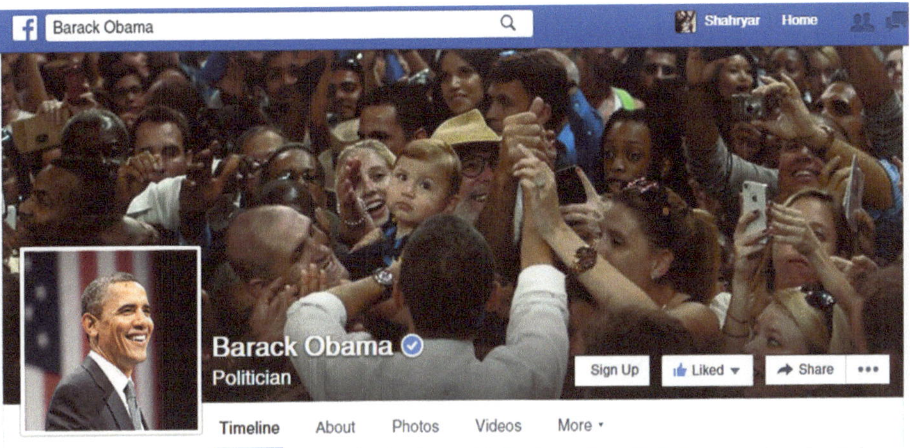

Bill Clinton
www.facebook.com/BillClinton

Bill Gates
www.facebook.com/BillGates

Mark Zuckerberg
www.facebook.com/Zuck

Rehan Allahwala
www.facebook.com/Rehan33

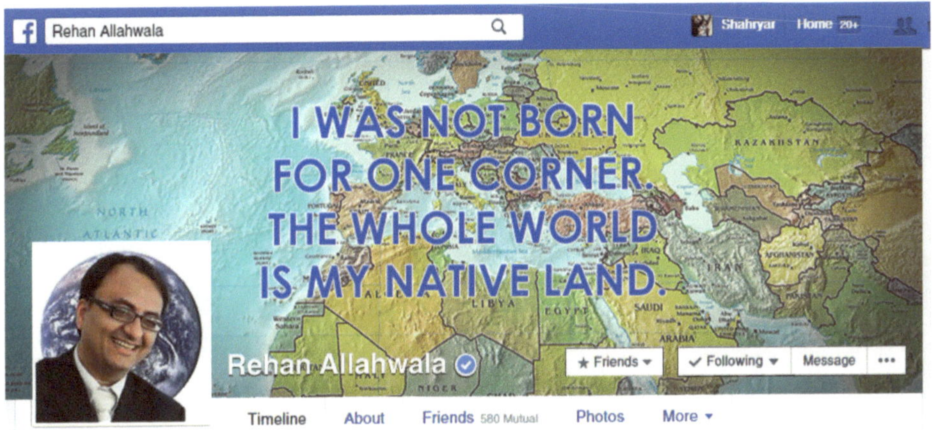

Paul Jones
www.facebook.com/MisterEntrepreneur

Richard Branson

www.facebook.com/RichardBranson

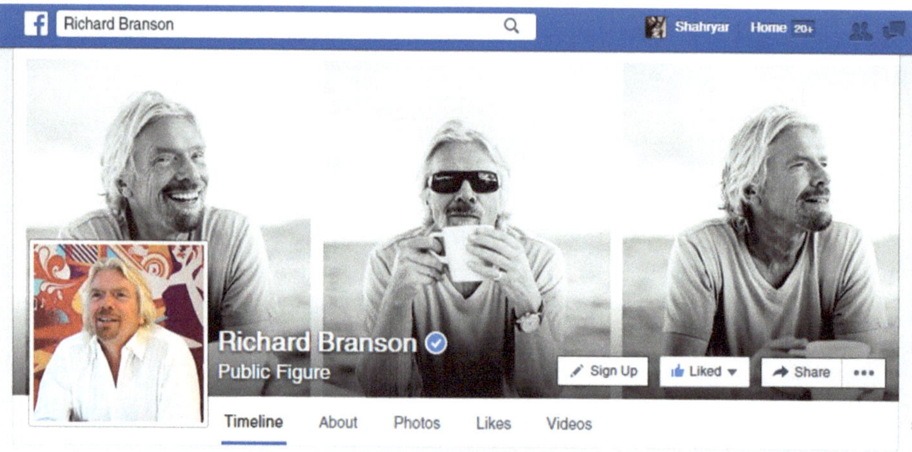

Suzanne Bowen

www.facebook.com/suzanne.m.bowen

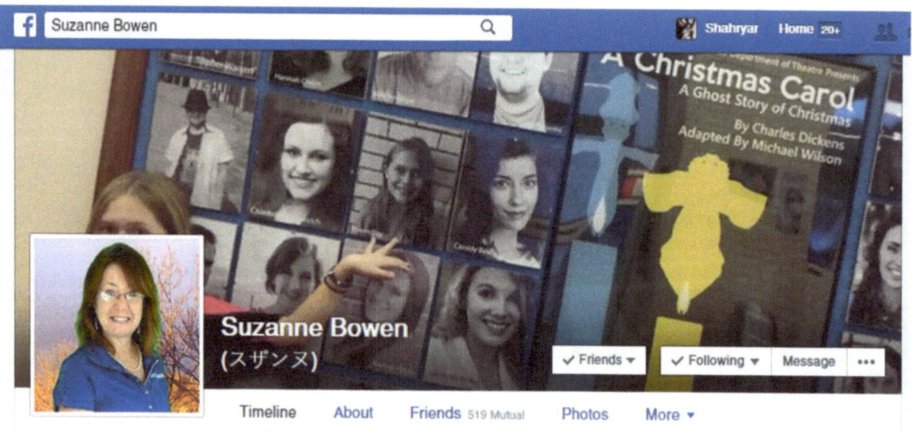

Munawwar Ikhlas Allahwala
www.facebook.com/MunawwarAllahwala

Zsuzsanna Fajcsak (Dr Zsu)
www.facebook.com/Fajcsak

Robert Scoble

www.facebook.com/RobertScoble

Steve Wozniak

www.facebook.com/SteveWoz

Qasim Ali Shah
www.facebook.com/Qasim.AliShah.50

Updated version of these people are also on:
rehanallahwala.com/mentors for people around the world.
rehanallahwala.com/pakimentors for people from Pakistan.

8. Searching For People On Facebook

Facebook now allows you to search for people from around the world, by different search categories. So you can use FB like Google. If you want to know people from one specific country, or you want to know people with a specific profession, or you want to know people with a specific interest **i.e.** Health, fitness, cycling, cars, technology, computers. Now you can search and connect with people by these specific criteria. Now it is easy to get first-hand information about your interest from these people. This way for sure you know you have common interest that you connect with and can become friends easily.

Here are some examples of what you can search on your search bar:

People who live in Albania

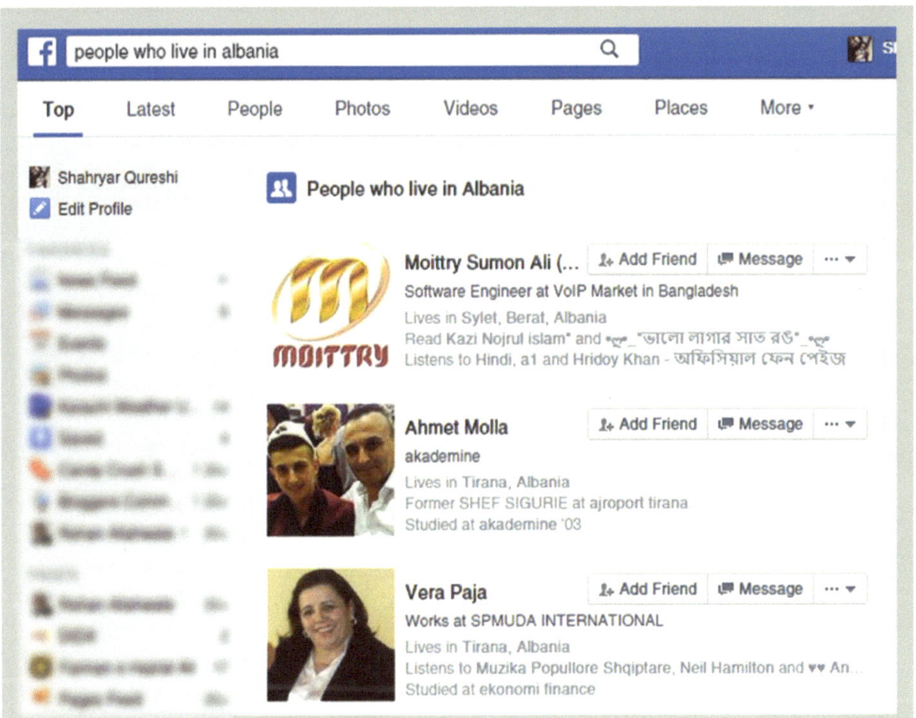

People who grew up in Germany

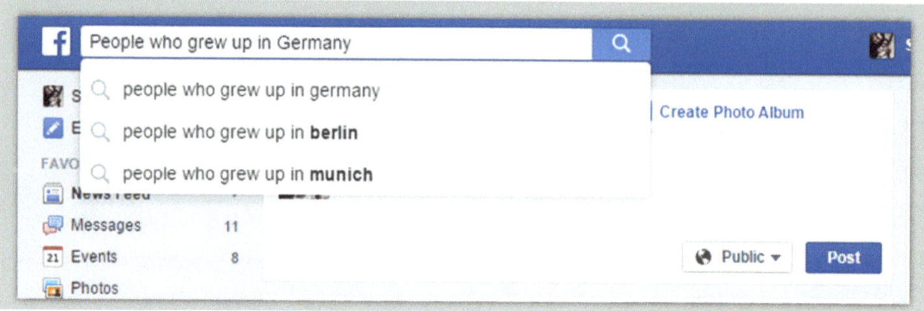

People who grew up in Germany
people who grew up in germany
people who grew up in berlin
people who grew up in munich
Create Photo Album
Messages 11
Events 8
Photos
Public
Post

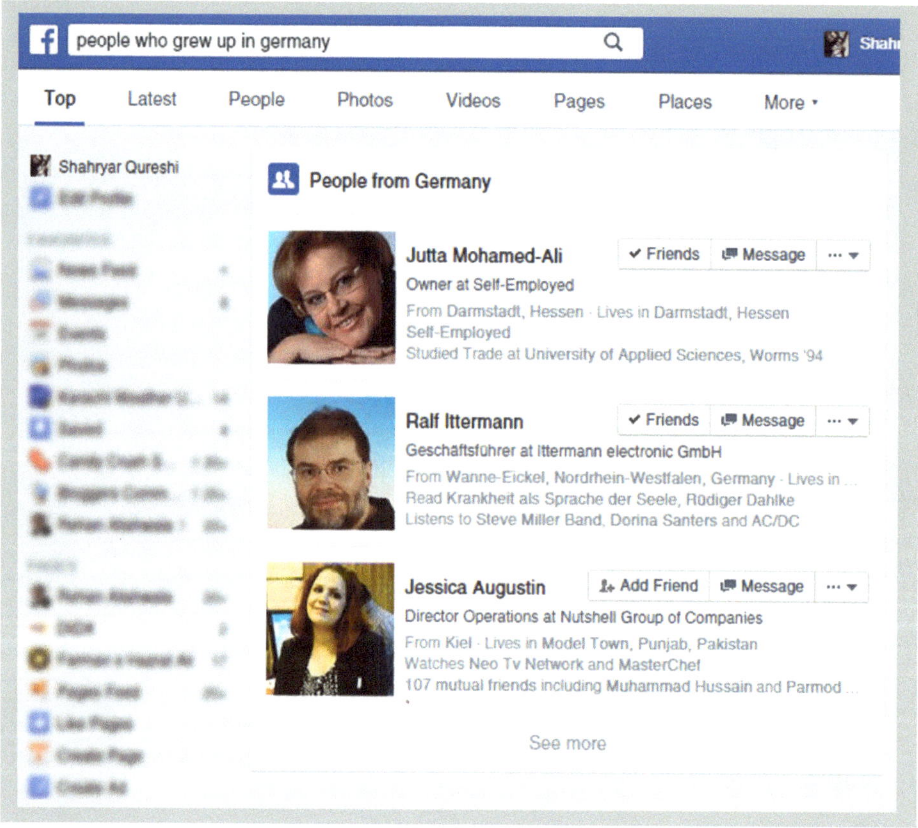

people who grew up in germany
Top
Latest
People
Photos
Videos
Pages
Places
More
Shahryar Qureshi
People from Germany
Jutta Mohamed-Ali
Friends
Message
Owner at Self-Employed
From Darmstadt, Hessen · Lives in Darmstadt, Hessen
Self-Employed
Studied Trade at University of Applied Sciences, Worms '94
Ralf Ittermann
Friends
Message
Geschäftsführer at Ittermann electronic GmbH
From Wanne-Eickel, Nordrhein-Westfalen, Germany · Lives in ...
Read Krankheit als Sprache der Seele, Rüdiger Dahlke
Listens to Steve Miller Band, Dorina Santers and AC/DC
Jessica Augustin
Add Friend
Message
Director Operations at Nutshell Group of Companies
From Kiel · Lives in Model Town, Punjab, Pakistan
Watches Neo Tv Network and MasterChef
107 mutual friends including Muhammad Hussain and Parmod ...
See more

Females who are interested in Purses

females who are interested in purses

Shah

Top Latest People Photos Videos Pages Places More

Shahryar Qureshi

Edit Profile

Women who like Purses

Emily Goodrich (E... Add Friend Message

Nevada High School Class of 2016

Single · Female · Interested in males

Likes Purses, Women in the military and 62 others

Worked at Interior Motives

Jade Jimbo Kennedy Add Friend Message

Wallace State Community College

· 28 years old

Likes Purses, Shriners Hospital for Children and 17 others

Studied Nursing at Bevill State Community College, Sumiton, A...

SHE HAD A LIVELY, PLAYFUL DISPOSITION, WHICH DELIGHTED IN ANYTHING RIDICULOUS

Amanda N. Patterson (Mandie) Message

General Manager at Stay at home Mommy & Wife!

Engaged to Andrew Jordan · Female

Likes Purses, Shoes in General and 1 other

Studies Associates Degree in Art, General Curriculum at Angelin...

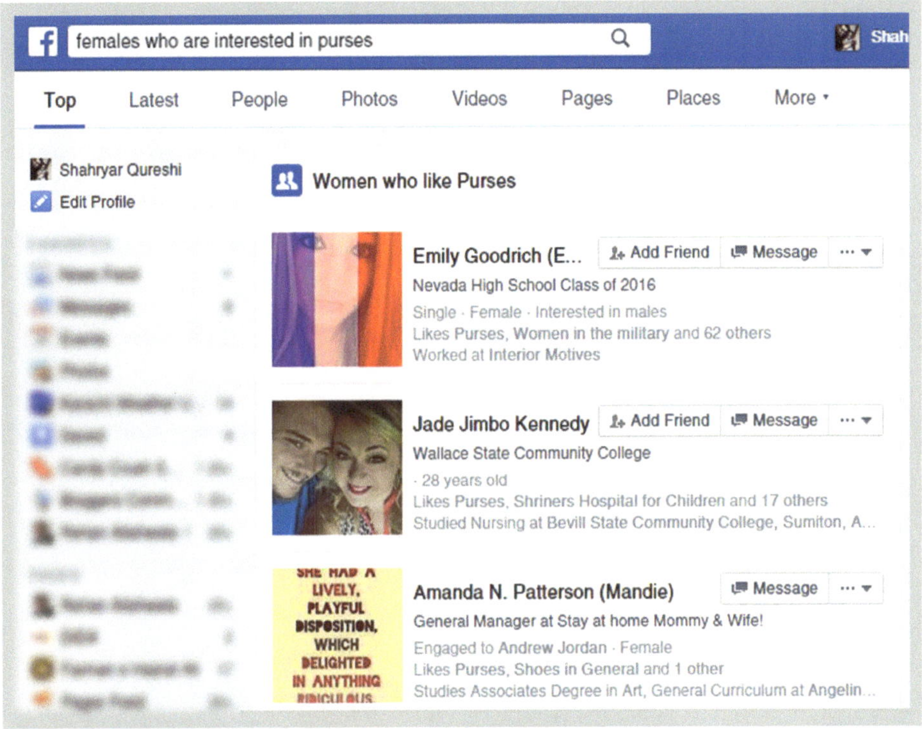

People who speak Russian and live in United States

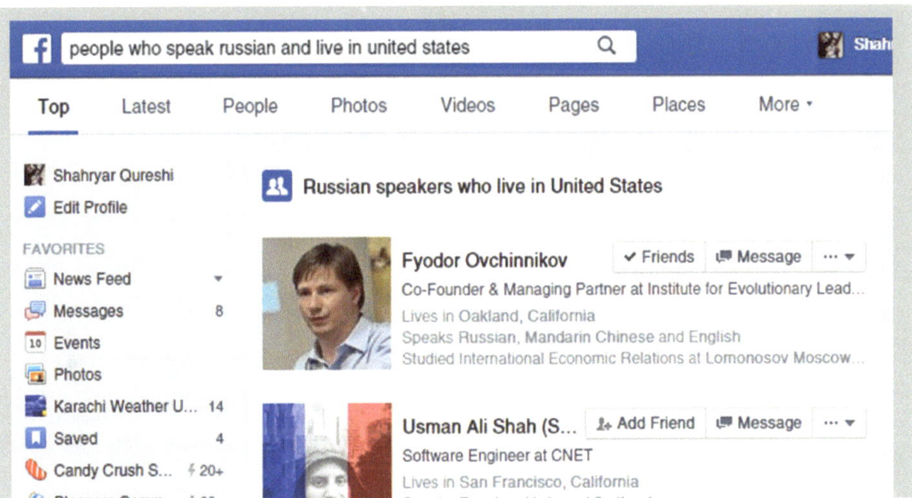

People who are engineers and live in Australia

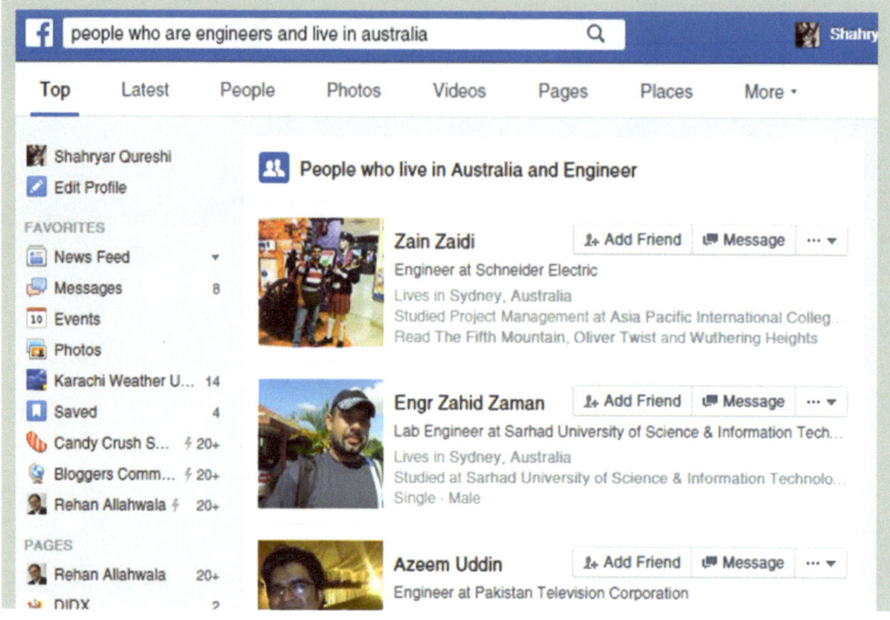

People who are singers and live in Hollywood

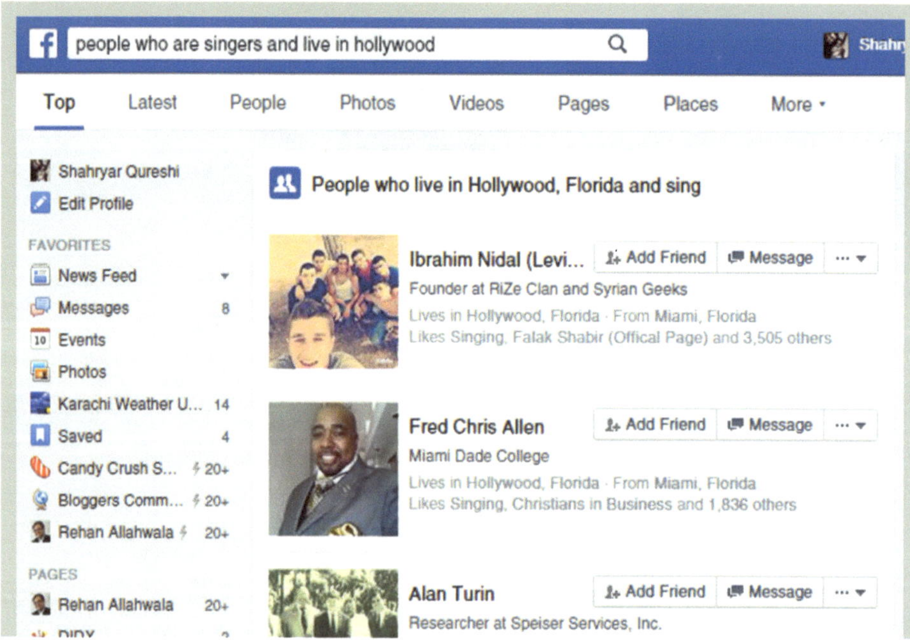

Some other search examples:

Males who are interested in Toyota Camry.
Friends I have from Germany.
People who work for Facebook.
People who work for Microsoft.
People who are looking for jobs.

This allows you to find different people from all walk of life and then friend them. They can help you to learn more about the specific subject what you are interested in.

9. Facebook Groups

Facebook groups are by far now the easiest way to make friends on Facebook. You can go and search for the interests that you have, join these groups, and read what people are posting.

When you join a group, I strongly recommend you do not post any thing to the group for a week, observe, observe and observe. Make sure you read what is going on. The maximum you can do is click like on the posts that you like. I recommend spending three days in the group before starting to comment. Then you can inbox to the members, start talking to them and explore if you both like to become friends.

WHY to be cautious:

A Facebook group is like a TRIBE, and you should not be speaking in a tribe among the members of the tribe, until you are sure what the rules of the tribe are. Every single group out there has a group administrator, group elders and these elders decide what goes in the group. Once you know exactly what is going on, you should speak according to the protocols of that group. Otherwise the admin may kick you out or even permanently block you from the group, for saying something that is out of line for that group.

Larger groups are less tolerant and more strict. Many spammers go and flood the groups with advertisements, which are not welcomed. If you are an advertiser, you will be blocked from these groups very quickly.

Screenshot of a Facebook Group. Group can be Public, Close or Secret

It is more than likely that the group or interest you have does not have the word you are seeking, so search a different synonym. If you still do not find, go ahead and create your own group on that subject. Add a few people who you know already of that interest and grow that network.

I highly encourage you to make a group about your town, city, country, business, hobby, building, sub-divisions.

Some Examples:
People living in Green Town, People from Dallas, I love PHP, I love Sewing, Sewing Club of Karachi.

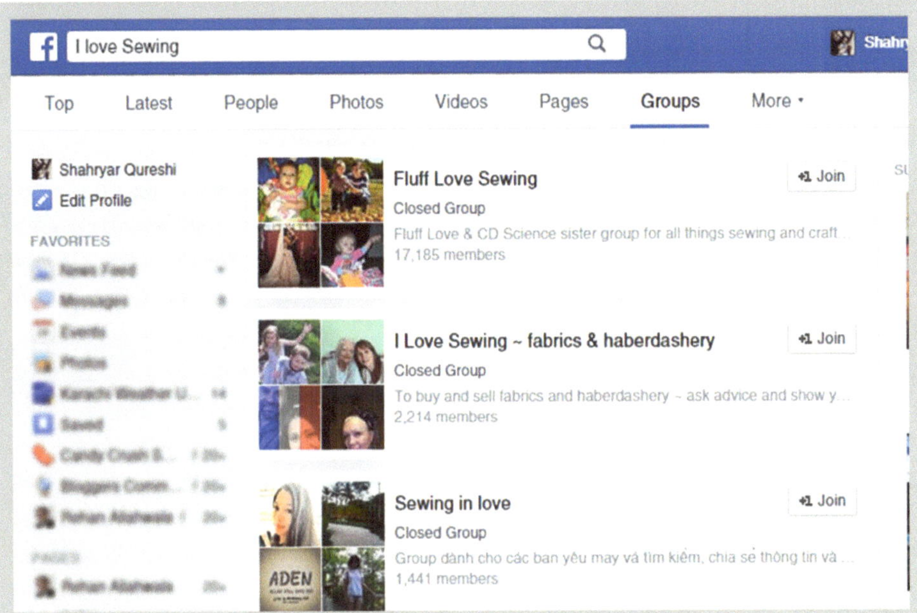

Joining a group will help you find more similar minded people eventually and help you grow that interest in you and in others.

10. Countries Where You Can Find People

Getting to know people and making friends from different parts of the world will help you to open your mind, grow as a human being, as you realize we are the same human beings all around the world.

Albania:

Albania was one of those countries that I did not know much about. So I posted a paid ad on Facebook, looking for friends from Albania, as the cost of doing that was minimum. To my surprise, it is a lovely country with low salary, highly educated people, with great infrastructure available to them. Unfortunately the lack of leadership is also there, and most of the people, even though they live in Europe, they make very low salary and are very depressed. Most people living there are Muslims, but very European kind of Muslims. I recommend making friends in Albania if you are Muslim, so you can understand them.

Macedonia:

Macedonia is also one of the ex Yugoslavian country. It is a small country with similar situation like Albania. Most people are not Muslims, but many of them are.

Note: You will find inter-religion marriages to be very common here.

Examples of countries where I recommend to reach out to and connect with people.

11. The 500-Mutual Friend Experiment

I was having lunch with my friends Mitch Carson and Ernesto Verdugo. I asked them, "**Ernesto - what can I do to get American people from America to talk to people in Pakistan? I think if these people would talk to each other, it would bring more love and peace among them?" He said, "Incentivise it and maybe give a gift or something, like a laptop."**

I thought about it, and a week later I announced the first laptop give away to anyone who will steal 500 of my Facebook friends.

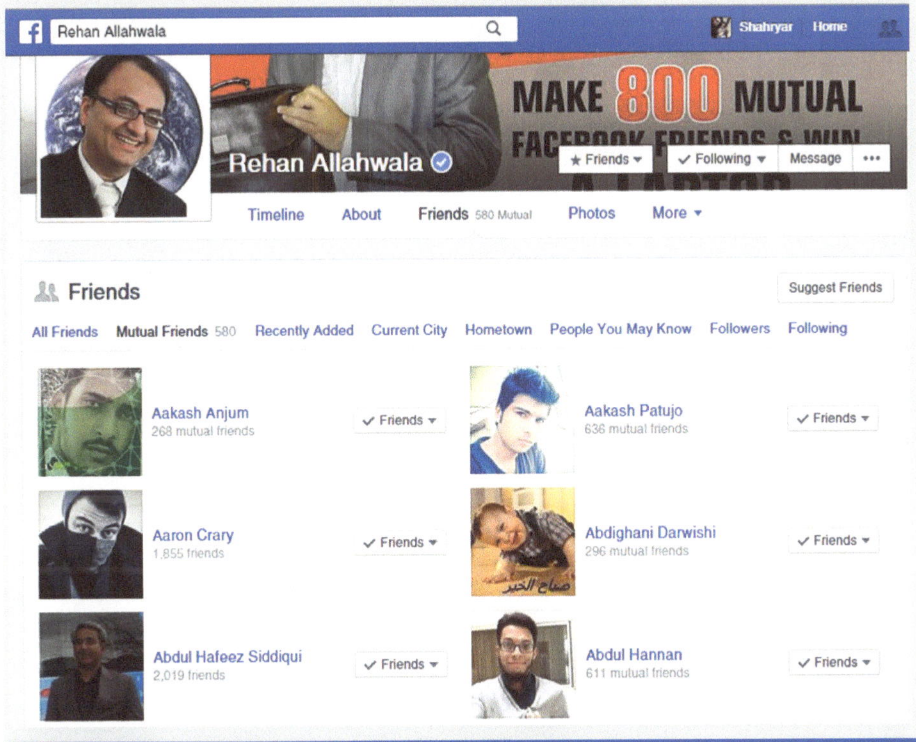

A Facebook user's over 500 Mutual Friends with Rehan Allahwala

YES, steal. What that means is that they have to go to my friends' list take 500 of my friends and make them their friends. If you are from Pakistan, you must make 500 Non Pakistani Friends. If you are from the US, you must make 500 Non US Friends, and so on.

This way I do not have to introduce people to each other and they have an incentive to make new friends.

This is just a small and earliest result of the laptop competition:

Shortly after the competition was launched, a young Pakistani man messaged me. He has managed to make 500 non-Pakistani mutual friends with me and he did this in only 2 weeks.

I asked him how he did that, his reply was: "I talked with every single one of them and convinced them that I am a good person. I gave them enough information about myself, so that they can trust me and that made them add me to their Facebook".

I was surprised that he could add all these people in only 2 weeks, as it took me ages to achieve this.

I was very happy to see that this boy's REAL journey has begun.

This journey will allow him to share exchange and bounce his ideas off of these very different 500 people around the globe. Interacting with them will allow him to think like a global citizen, being considerate and appreciative about the difference among all of us. This will increase his chances for success.

The complete story of the 500 Mutual Friends Experiment will be coming soon in our separate book.

12. Success Stories Of Those Who Expanded Their Friends On FB

Zsuzsanna Fajcsak

/Fajcsak

When I met Rehan at the end of 2010, I had 95 friends on Facebook. He convinced me it will be very good for me to start connecting with people and adding them Rehan made the mind-set shift in my head, that I should give out knowledge on Facebook and teach others for free as social contribution.

This way my work will be visible and I create a social proof of my work, as what others comment is a very valuable piece of my work. I spent 4-5 hours on fb initially talking to people and helping them with their health issues. I made 1000 people in 4 months, then 3000, then reached 5000 friends in 2 years, I Think. How did adding people helped me?

My work began to be visible and I received more and more invitations for sessions. I was very happy that I could reach out to so many people and help them. Moving to Pakistan and making to be known in only in 4 months, well that's the power of Facebook. Adding people helped me to realize that not everyone is bad. Facebook surprised me so many times with getting to know really nice people. And if we just reach out and be open minded talk nicely everyone will reflect that.

A lot of people who need help don't go to get help, but they are willing to talk to professionals on Facebook. I can only encourage more professionals to spend a little CSR time on Facebook and mentor and help the youth as many of them are lost. Many adult are depressed too, exactly 80% of people are depressed on this world and are living without help, as they are ashamed. Then the "Follow function came and this allowed to expand my audience even further. After 5.5 years I have 18,000 followers on Facebook enjoying the positive posts and getting help for happier healthier living.

Yollana Shore

Since Rehan Allahwala introduced me to his friends, I have had hundreds of friend requests. On the first day, I had many conversations with people and I asked them all one question - if you had one message for the world, what would it be? I want to share with you what some of the good people of Pakistan said to me... :)

Susan Webb

ADD ME ~ can two words typed over the internet from a total stranger on the other side of the planet change your life? Can five letters of a symbolic code change one's perspective? The outcome ironic in many ways. The choice in hindsight immediate. The click of the mouse on the "add as friend" as historic a moment as I will ever experience.

As I scrolled the names of comments on a fish game on Facebook, one stood out. The profile picture: Freddy Krueger; the name Kabeer Khan. The irony of this persona did not escape me. In the weeks to come there were gifts of fish given, aquariums visited. For every gift, I received a thoughtful acknowledgment.

After one particularly heartfelt thank you, I visited his profile. His profile statement ~ **"I am hemophiliac from birth at Fatimid. You must help us."** It listed the country as Pakistan.

I then followed a link to a group. This explained that Fatimid was a hospital for poor people with blood disorders. I sat and read that statement over and over. **"You must help us!"**

The desperation even at this moment brings tears to my eyes. No question, no option. Pure desperation. The choice so clear in this young man's mind. This statement so deliberate became a mantra in my head. What could make a situation so urgent, to demand?

After a day or so of pondering these questions, and getting up the nerve to speak to someone that would put a serial killer on their profile, I sent a chat message. I said: "Hello." I asked: "How I could help?"

I offered to make an American page for Fatimid to show support. In minutes, I was an Admin on his group. Accepted without question as someone who was sincere. At this moment, my life changed forever. My perception of the world, the internet, what possibilities were out there that I have never considered, everything changed. In one moment two people's lives would never be the same. In seemingly one moment a bond was formed. A thread of love, hope, and caring was without knowing about to sew the most beautiful tapestry imaginable.

What I found was an incredible humanity fostered by desperation. A family locked in a daily struggle. Not for what we would consider basic needs, but for a need at an even more vital level. Blood. Broken down even further one component of blood. A component that all four brothers in this family need from outside sources to survive. In a country that faces struggles on a daily basis that we in the west can only see in our nightmares.

I found a mother. A loving, caring mother, who has been through struggles I think I will never know. Brought up 4 productive, loving boys. Boys who do not wallow in self pity. Boys who work hard to make their parents proud, and to make Pakistan a better place. I found a father who worked 2 jobs to put his children through school. I found without knowing at the time, a family. sons I never dreamt I would have. A sister so dear I long to meet. I found people I miss every day, without ever knowing and in the process I found ME.

Mohammad Layeeque

/layeeque86

Thank you Rehan Allahwala, Thank you Mark Zuckerberg and Thanks a lot all Friends and Followers. Two years back when I was in Mumbai working with a Bank and was trying to connect with Corporates and Entrepreneurs in abroad, by random search I found him on LinkedIn. Seeing unique surname, wanted to know more about him and searched on Google, YouTube and other search engines, watched his videos and found him a person with amazing ideas.

Finally I search his Facebook ID, and send friend request, like on LinkedIn.

Soon got connected with him on fb and he became #FirstPerson from #Pakistan I connected and talked a lot.

I thanked him for accepting friend request and he asked "should I introduce more people to you on Facebook" and I agreed, and from that day my Facebook social life began.

I remember that time I had arpx. 500 friends on Facebook and most of them were my colleagues, school and college friends (from my country India only).

He kept introducing me to new people almost from Pakistan first and later from other countries. I loved talking to them, knowing them and adding on Facebook.

There was a day in 2008 when I was in Dubai and unknowingly met a Pakistani national in a barber shop and thought I gonna be killed today and somehow managed to leave that place soon without revealing that I was an Indian.

This fear was with me for them because of what I and people of both these nations learnt from media and people around us, that is all negative and hateful things about neighbour's country.

From that time when I was afraid of them and I had only 500 friends on Facebook, till now where Facebook friends limit - 5000 is full and connected with more 2700 people as #Followers and I believe out of 7,700 people on my Facebook, 4000 are alone from #Pakistan, who accepted and trusted me as an Indian, with #IndianFlag on my profile pic and the name #India with my profile name.

I'm thankful to almighty Allah for making this happen to me, Thanks a lot to Rehan Allahwala for adding me on social media and helping to connect people from Pakistan, India and all around the world and Thank you soo much all Friends and Followers for your trust and being so nice to me. .. Thanks a lot for reading.

DIDx.net is the trademark of Super Technologies Inc.

WWW.DIDX.NET

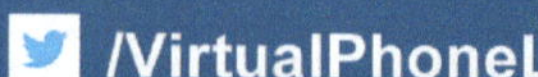
/VirtualPhoneL

www.ingramcontent.com/pod-product-compliance
Lightning Source LLC
Chambersburg PA
CBHW041241200526
45159CB00029B/96

* 9 7 8 1 5 2 3 7 2 5 1 4 4 *